CAREERS IN
ENGINEERING
STRUCTURAL ENGINEER

EVERYWHERE YOU LOOK, YOU WITNESS the work of structural engineers. These professionals are responsible for ensuring that every structure is safe and sound, whether it is a building, vehicle, or part of infrastructure. They study how to make buildings withstand the onslaught of earthquakes, hurricanes, extreme weather, and other natural forces. They improve the way structures are built, help minimize the impact of construction on our planet, introduce new and stronger materials, and find the best ways to utilize sustainable resources.

Structural engineers are involved in every step of the building process. They draw up designs from scratch and collaborate with architects and other kinds of engineers to create buildings that can fulfill their intended use. Structural engineers design the framework of large structures like skyscrapers and bridges to make them capable of supporting their own weight while resisting the forces of weather and traffic. They design specific architectural components like beams, columns, foundations, and floors that need to be structurally sound. They draw on their expertise with various materials to choose the most appropriate materials for each job.

Structural engineers often specialize in the types of structures they design and may work on projects ranging from residential homes to nuclear power plants. They also breathe new life into old buildings, renovating or transforming them to serve completely new purposes. In some cases, they inspect old buildings and direct their demolition. If a structure fails, they may be called upon to investigate the cause. Regardless of the size or scope of the project, their main focus is always on the safety and feasibility of the design.

Although structural engineering is closely associated with the construction of buildings, the professionals are also involved in the design of machinery, medical equipment, and vehicles. Their skills and expertise are needed wherever structural integrity affects functioning and safety.

It takes considerable knowledge and skills to do the work of a structural engineer. Because of the safety issues involved, structural engineers are trained to strict standards. Most structural engineers start their careers with a bachelor's degree in civil, mechanical, or aerospace engineering, with specialized courses covering the basic concepts of structural engineering. Although a bachelor's degree is enough to qualify for most entry-level jobs, a master's degree in structural engineering is needed to advance to more senior-level positions.

The educational path is intense, but once qualified, new structural engineers become highly sought-after professionals. Engineering projects are in high gear, and opportunities are everywhere. Structural engineering jobs can be found in small consulting firms and large multinational corporations with offices around the world. There are opportunities for travel and working overseas, since the skills needed for structural engineering are the same anywhere in the world.

Structural engineering is a hugely satisfying profession with both tangible and intangible rewards. Because the demand is currently exceeding supply, structural engineers are enjoying good pay that continues to get even better. Employers are attracting qualified candidates with signing bonuses and a bucketful of exceptional benefits. There is also a great deal of variety, creative satisfaction, and the chance to help shape a better world. Structural engineers are highly respected for their contributions to society. It is a career you can be proud of.

YOUR ASSIGNMENT

STUDY HARD IN ALL OF YOUR HIGH SCHOOL CLASSES and fill your schedule with the highest level courses offered. This is the best way to prepare for the rigors of college. Take all the math courses available. Advanced math skills are extremely important in this field, so consider taking some college-level classes, especially AP (Advanced Placement) Calculus if is available at your school. Physics is another important class that should be included in your curriculum.

Computer proficiency is essential. Structural engineers use computer-aided design (CAD) systems more than any other tool. Take a computer class in school, join a computer club, or take some (free) courses online.

Learn as much as you can about the structural engineering career. For inspiration, read books and watch documentaries about famous engineers and special projects. Look for lectures and workshops presented by architects and engineers that you could attend. Ask your guidance counselor to invite a structural engineer to speak on career day. Call some structural engineering companies and ask if you can shadow one of their engineers for a day or two. This will give you insight like nothing else can. Structural engineers love sharing their very rewarding field with anyone who is genuinely interested and enthusiastic. It will help you understand exactly what goes on in the day-to-day activities, and you might even find someone willing to be your mentor.

Look for opportunities to intern for structural engineers. You do not necessarily have to be in college to participate. There are some structural engineering internships available to high school students – some even offer pay! Ask your school counselor to help you find them or search through online job sites. You can also offer to intern for local engineering firms.

Experiment on your own. Build things, fix things, take things apart. Figure out what works and what does not, and why. Start with toothpick bridges and trusses and work your way up. You can find ideas for fun projects on the internet.

HISTORY OF THE CAREER

STRUCTURAL ENGINEERING IS AMONG the oldest types of engineering. Its history spans dozens of millennia, from the Neanderthal shelters constructed of mammoth bones and vines, to the International Space Station.

The first structural engineer known by name was Imhotep, who was responsible for building the step pyramid for Egyptian Pharaoh Djoser in the 27th century BC. The Pyramids are marvels of engineering because their structural form is inherently stable and can be almost infinitely scaled.

The knowledge base that forms the foundation of structural engineering in the Western world is rooted in the work of Greek engineer, Archimedes (250 BC). His studies and publications focused on the mathematical concepts of calculus and geometry, which could be used to determine centers of gravity. Among the ancient civilizations, the Romans were also known for their superior structural engineering. They were particularly skilled at using masonry and concrete to build large structures, like aqueducts, huge public bath complexes, architectural columns, defensive walls, lighthouses, and harbors. The methods used to build these structures – many of which are still standing today – were described in *De Architectura*, written by military engineer Vitruvius.

It was not until the Renaissance that the physical sciences underlying structural engineering began to emerge. Leonardo da Vinci, a self-taught engineer, produced many engineering designs based on scientific observations and rigor. At the time, his designs were often ridiculed as fanciful and dismissed, but have since been judged to be feasible and structurally valid.

The progress of structural engineering picked up steam in the 17th century with the help of several renowned scientists. Galileo Galilei's book, *Two New Sciences,* established a scientific approach to structural engineering and introduced structural analysis. It examined the failure of simple structures, discussed the strength of materials, and referred to the science of the motion of objects.

Sir Isaac Newton followed up on the discussion with his publication, *Philosophiae Naturalis Principia Mathematica.* In the book, Newton's Laws of Motion laid the foundation for fundamental laws governing structures.

English architect, Robert Hooke, added to scientific understanding with his observations of elasticity of materials and the effect of load.

The study of materials and the development of structural analysis continued throughout the 19th century, with particular emphasis on the elastic behaviors of structures, compression under load, and methods of reinforcement in both materials and structures. New materials were introduced, but reinvented materials proved to have equal impact.

Cement, which was used throughout ancient civilizations, went through several transformations. In 1824, Portland cement was patented by a bricklayer named Joseph Aspdin. It was the first artificially manufactured material that became masonry mortar when mixed with water and sand. After it hardened, it closely resembled limestone from the Isle of Portland, which had long been the preferred choice for building churches, mansions, and palaces. It turned out that the new Portland cement was even more durable than its famed namesake. The significance of Portland cement, however, was in its economy. Because it was made of cheap and commonly available materials, concrete construction, which was previously too expensive for wide use, was now economically viable. Soon after the invention of Portland cement, reinforced concrete made its debut. Using steel mesh and steel reinforcement bars located in the areas of tension within structures was a major step towards building structures that would not collapse from movement or load.

Iron was another raw material that underwent transformations in the 19th century. Cast iron gradually replaced the traditional wrought iron. It was embraced by engineers who were seeking ways to build structures that would not burn easily. At first, it was used mainly to produce columns in mills, and in fact, the first fire-proofed building was the Belper North Mill, in England, built by William Strutt, a pioneer in a number of innovative techniques including using a "fire-proof" iron-frame, instead of timber. It became a blueprint for future mill construction and in addition was a template for modern skyscraper engineering. Eventually, cast iron was replaced with the first major uses of steel.

At the dawn of the 20th century, the profession of structural engineer was really taking shape. High-rise construction was now possible thanks to the advancements of the previous century. Problems associated with stress in its various forms were addressed with inventions like prestressed concrete, more complex mathematical analysis, and newly designed structural systems that could withstand lateral forces.

The underlying goal in the 20th century was to prevent structural failure. In fact, some of the most significant advancements in structural engineering resulted from the study of structural failures. For example, the Tacoma Narrows Bridge (known as Galloping Gertie) collapsed in 1940. The collapse and its causes are studied by every structural engineering student as the classic example of resonance. The bridge was an elegant and graceful design that looked like strings of spaghetti strung between two towers. When the winds passed through it, the bridge began to rise and fall in vertical waves. The oscillation was of such magnitude that the bridge eventually tore apart and plunged 190 feet into the Puget Sound. The research that followed the collapse led to a greater understanding of wind-structure interactions, and wind-tunnel testing of bridge designs became mandatory.

Another noteworthy failure was that of the world's first commercial airliner that crashed in 1954. Lengthy and meticulous investigations led to the discovery that metal fatigue at the corners of the windows was the culprit. At that time, the corners were square, which led to concentrations of stress. The continual pressurization and depressurization emphasized the stress until ending in tragic failure. The research into the failures led to a greater understanding of fatigue loading of airframes, and all subsequent airliners have incorporated rounded door and window corners.

Throughout the late 20th century, the modern computer became the structural engineer's most valuable tool. It made it possible to apply finite element analysis to structural analysis and design. Sophisticated software programs allowed engineers to accurately predict the stresses in complex structures. Many modern structures could not have been designed and successfully built without the use of computational analysis. There is now a tremendous body of knowledge available in structural engineering and a growing range of different structures. The increasing complexity of those structures has led more and more structural engineers to focus on particular areas of specialization.

The names of the earliest practitioners of structural engineering are largely unknown. We do not know who designed the Hanging Gardens of Babylon or the Parthenon. We can only marvel at their existence. We do recognize the prominent structural engineers of the Eiffel Tower, Statue of Liberty, Gateway Arch, and James Webb Space Telescope. The structural engineers of tomorrow are sure to bring us more wonders than we can imagine.

WHERE YOU WILL WORK

STRUCTURAL ENGINEERS TYPICALLY WORK in metropolitan areas where there is a high demand for their services. But there are also smaller firms and local government agencies that hire these professionals in small and medium sized cities with growing populations. The majority of structural engineers are employed by private engineering consultancy firms. Some work in civil service employment at the federal, state and local levels, but in recent years, most government work is contracted out to private firms.

Aside from structural engineering firms, other major employers of structural engineers include:

- Railroads
- Oil and gas companies
- Public utility companies
- The Armed Forces
- Aeronautic manufacturers
- Civil engineering contractors
- Automotive manufacturers
- Medical research firms

There are literally thousands of structural engineering firms operating in the US. They come in all types and sizes, with varying areas of special expertise. Smaller firms offer more autonomy and the opportunity to explore different aspects of the work. Large firms offer professional prestige, opportunities for career growth, and excellent compensation packages. For the new engineer who wants to do amazing things, bigger firms are also more likely to have large-scale award-winning projects. You can see and do amazing things at these companies. The five largest companies hiring structural engineers are Boeing, URS, Jacobs Engineering, HDR, and Burns & McDonnell.

Boeing is a household name because the Boeing Company is the largest and oldest manufacturer of American military and commercial jetliners. Its many structural engineers continually focus on the structural integrity of facilities and products. Structural engineers who land a job at Boeing have the opportunity to work with some of the world's most brilliant and talented designers.

The URS Corporation is a leading design, engineering, and construction management firm. The firm is constantly recruiting structural engineers to work in its offices throughout the Asia Pacific, Europe, the Americas, and Africa. Although the name may not sound familiar to anyone outside the industry, the company has an excellent reputation for top-notch design and construction projects of all sizes. Major industrial and commercial clients around the world seek out its structural engineering services.

The Jacobs Engineering Group is another leading design, construction, consulting, and project management firm that operates in various locations around the world. Because operations are scattered throughout many countries, it represents a good choice for those who want the opportunity to travel the world.

HDR is on the same playing field as Jacobs Engineering, with its high-level design, construction, project management, and consulting services. It is best known for landing high-profile industrial, commercial, and government contracts. This is an international operation with offices located in Europe, Asia, Australia, the Americas, and the Middle East. Structural engineers working for HDR are exposed to a wide variety of projects, including the construction of roadways, bridges, buildings, waterways, and dams.

Burns & McDonnell has the reputation for being one of America's best engineering companies. The firm serves industrial, commercial, and government clients, and has high-level expertise in the aviation, utilities, defense, and environmental markets. There is also a new department in the firm dedicated to serving the healthcare industry. One of the more interesting aspects of this big company is that it is wholly owned by its employees.

Work Environment

Although structural engineering is usually office-based work, these professionals also work in a variety of locations and conditions. Those involved in construction need to visit sites to check dimensions, site locations, available space, height restrictions, and other details. This information is mostly gathered in the beginning stages, then it is back to the office to work on computers and develop plans. A final site visit is usually carried out to check that what is proposed can actually be built. This same routine applies to projects in industries other than construction, such as aeronautics. Some jobs may also require frequent travel. Depending on the size and location of the project, overseas trips may be a standard feature.

Work Schedules

Most structural engineers work full time on a standard 9 to 5 timetable with weekends and common holidays free. However, there may be a need to work overtime and visit sites on very short notice. If a building or structure is in danger of collapsing, for example, structural engineers will need to travel to the site immediately. There are also times when external conditions, such as bad weather or environmental noise, limit the hours that can be worked near populated areas. Though full-time work is the norm, it is often possible to limit hours to part-time work.

THE WORK YOU WILL DO

STRUCTURAL ENGINEERS ARE INVOLVED in the design of structures with a focus on ensuring those structures can withstand the stresses and pressures of the environment and usage over an extended period of time. They analyze plans, design components, select materials, and determine project viability. They are primarily concerned with safety, technical, economic, and environmental issues, but they also take into consideration any aesthetic and social factors.

Structural engineering is vital to the construction process, whether the structure is a car, a ship, an office building, or a space station. All structures will experience various stresses and strains over time. It is the structural engineer's responsibility to make sure the materials, components, and designs used to create the structures will keep them safe, sturdy, and durable. In short, they make sure buildings do not fall down and car crashes are survivable.

Structural engineers use advanced mathematical calculations and technical knowledge to analyze the many possible physical forces at work on structures. Without the input of a structural engineer, buildings and bridges could twist, bend, or vibrate beyond their limits, causing structural damage and putting lives at risk. To avoid the worst outcomes, structural engineers look at many factors, from the type of soil to the possibility of withstanding a blast explosion. They take into account the weather, including seasonal and possibly extreme temperature changes, wind, rain, snow, and drought conditions. They make sure structures can handle earthquakes, heavy traffic, gravity, collisions, and the size of the structure itself. They consider the number of stress cycles over the lifetime of bridges or aircraft that eventually cause fatigue.

Structural engineers also choose the most appropriate materials for each project. Materials need to be durable, strong, flexible, and unlikely to deteriorate over the expected lifetime of the structure. Structural engineers have expert knowledge of the features of steel, concrete, wood, masonry, aluminum, composites, and more.

What Kinds of Projects?

Structural engineers work on a wide variety of projects. The structures most commonly associated with structural engineers include buildings, stadiums, towers, bridges, dams, and roadways. The buildings they design are often large, commercial projects like hospitals, schools, apartment buildings, office buildings, and government facilities. In the construction industry, they work closely with commercial architects, civil engineers, mechanical engineers, electrical engineers, quantity surveyors, and construction foremen.

The majority of structural engineers are employed in the construction industry, often on civil engineering and public works projects, but that is certainly not the only kind of work they do. They play a major role in anything that will be constructed. For example, they design oil rigs (both offshore and onshore), space satellites, aircraft, ships, trucks, pipelines, waterways, medical diagnostic tools, and (tiny) nano lasers.

Not all structural engineers work on massive and impressive projects. Many focus on residential structures. The very homes we all live in are directly affected by their work. Their role is essential in making homes safe, especially in areas that are known to be subject to nature's fury. For example, what if it rains hard on homes built on steep hillsides? What if an earthquake shakes a condo in California? What happens when a hurricane sweeps through the South? It is the structural engineer who knows how to make sure homes in these situations are left standing when the danger passes. Even if a particular home has not been seen or designed by a structural engineer, their work is at play. Structural engineers heavily influence the safety regulations that govern what materials contractors can use and the building codes that ensure the safety of the people who will live in the homes.

Structural engineers also work on existing structures. In the construction industry, they are often called in to advise on repairs, or help design conversions and extensions. They also inspect older structures, which may not have been built to current codes, to determine whether they are safe. Their designs may call for major repairs, refurbishment, or repurposing. In some cases, they are involved in the demolition or dismantling of a structure that needs to be replaced.

Some structural engineers investigate failure, such as collapses or major foundation cracks, as well as performance problems. This is a highly specialized area known as forensic structural engineering that requires the usual expertise in structural engineering plus knowledge of legal procedures. The primary goal is to determine the probable cause of failure. This may mean examining the materials and components involved, determining whether an illegal or improper activity was involved, collecting physical evidence, and comparing witness statements to the evidence. They may also assess the damage and cost of repairs or replacement.

Job Tasks

The tasks involved in structural engineering are varied and complex. Actual day-to-day duties depend on the type of project or industry. However, all structural engineers spend the bulk of their time on the computer. They use specialized software known as Building Information Modeling or BIM that allows them to do 3D modeling and information management. In other words, it can be used to handle the elements of an entire project. Do not misunderstand the word "building." In this context, building means any 3-dimensional object and therefore, it is useful for any structural engineering in or outside the construction industry.

In addition to general design tasks, structural engineers use their computers to do any of the following:

- Analyze blueprints, maps, reports, topography, and geological data
- Determine the best materials for the job and estimate the quantities and cost
- Compute load and grade requirements, water flow rates, and material stress factors
- Study traffic patterns or environmental conditions to identify potential problems that might affect the project
- Use computer models to predict how structures will react to the weather
- Work out ways to improve energy efficiency
- Write and present progress reports to clients and senior managers

Away from the computer, structural engineers meet regularly with clients, architects, and engineers involved in each project. They inspect project sites to monitor progress and ensure the project is being built according to design specifications. Structural engineers may also be required to assist government bodies in their own inspections relating to the project. They make sure health, safety, and other legal regulations are maintained. They may be called upon to assess commercial buildings or other structures prior to their sale or occupancy.

In the construction industry, the structural engineer will often work closely with the architect before the shovel hits the dirt. Then when construction begins, they turn their attention to the general contractor to ensure that the proper materials are used and the design is completed as planned in the initial phase.

The architect is usually the lead designer on buildings, with a structural engineer employed as a consultant. However, this hierarchy may change depending on the complexity of the structure. Many structures, like housing and multi-story office buildings and condo complexes are structurally simple. These are most likely led by the architect. Other structures may be heavily dependent on their form for their strength. The structural engineer is going to have more significant input on tensile structures (those made with membranes), shells, or gridshells (shells with double curvature). The Sydney Opera House is a good example of a project led by structural engineers. Basically, architects are concerned with aesthetics and base their designs on the intended appearance, shape, size, and use of a building. Structural engineers solve the technical problems so that the architect's vision can be achieved.

Some structural engineers are more involved in the construction process and therefore, spend considerable time on site. They may be responsible for preparing budgets, ordering and overseeing the delivery of materials, managing on-site labor, and supervising design and technical teams. These structural engineers often install a trailer on site to serve as an office for the duration of the project.

Specializations

Many structural engineers advance their careers by becoming project managers or starting their own consulting firms. A few will move into the university environment. Many more choose to specialize in a specific area of structural engineering that will make them more valuable in fields that are most in demand. For instance, some may choose to work with buildings constructed from one particular sort of material, such as green construction's novel materials. Some of the other common types of specialization include:

- Earthquake engineering
- Medical equipment
- Nanoscale materials
- Aerospace
- Mechanical structures
- Marine
- Automotive

STORIES OF STRUCTURAL ENGINEERS

I Work on Public Works Projects

"I chose my career after a natural disaster showed me how important it is that we have safe structures. I was in the ninth grade when the Loma Prieta earthquake hit the Bay Area. It was a stunning event, a 6.9 quake that killed 63 people, pancaked freeways, and caused the collapse of the Bay Bridge that had stood strong since the early 1930s. News stories with video footage played over and over on local stations. For months I watched the progress and learned why different structures were damaged and why some could be rebuilt while others had to be demolished. I also learned that the people figuring all this out were structural engineers. By the time I was ready to fill out my college applications I knew that is what I wanted to be, too.

Because I work in the public sector, there is a lot more paperwork than I expected. The bigger the project, the more time I spend in meetings, writing reports, going over budgets, and dealing with construction problems and deadlines. It can feel overwhelming and it is sometimes frustrating to be sidetracked from what I really want to do, which is designing buildings that will stand up to extreme forces. But in the end, when the structure is finished, I get great pleasure in seeing the connection between my work and the safety of my community."

I Work With High-End Home Owners

"I am an independent structural engineering consultant. After working for an international engineering firm for five years, I realized I could have a lot more freedom (and more money) by going solo. I was good at networking and had a lot of contacts from my college days that I could tap for business to get started. It was a risk, but I was surprised when I had enough business within a month to make ends meet. A year later, I moved to Costa Rica. My client base is still in Seattle and I rarely meet a client face-to-face. All my meetings are on conducted online.

I do structural designing for new homes and rehabs. Client homes often were built before regulations changed, and many are essentially unsafe. The neighborhoods where my clients live are hilly and generally wet. That's not a good combination. Then there are homeowners who want the 'open space' concept and don't know what structural components need to be preserved or added. I am able to make the necessary calculations from a thousand miles away, earning very good money and spending very little."

I Work on Historic Buildings

"For some structural engineers, the workday is all about doing calculations on the computer. The work is much more varied in my job. I usually do some calculations every day, but I spend more time drawing sketches, analyzing computer models, looking up records of old buildings, and visiting sites with my team. It's the variety in the workday that appeals to me.

Some engineers like the theoretical side and are happy to spend their time solving problems with complex math and physics skills. But there is a softer real-world side, and I think of this as a creative profession. When I'm first called out to inspect a hundred-year-old building, I visualize what it would look like if it were updated aesthetically and how I could make that happen while ensuring it could stand for another hundred years. By working as part of a team that includes architects, mechanical engineers, and electrical engineers, I can make sure my structural changes don't cause problems during construction and don't get in the way of the aesthetics. The best way to save a beautiful old building is to create a future vision as a team."

PERSONAL QUALITIES

STRUCTURAL ENGINEERS ARE WELL-SCHOOLED in higher math concepts and physics. Though every one of these engineers has the training to do the job, there are additional unique qualities shared by the most successful among them.

Technology, like 3D modeling software, is in every structural engineer's toolkit, but it is not enough to know how to use it. It takes creativity to use it to design the structures and systems that improve our many roadways, bridges, high-rises, and vehicles big and small. The best structural engineers are creative people who think outside the box to come up with new ways to ensure that structures, spaces, and vehicles are safer and more efficient, as well as cost-effective and beautiful.

Structural engineers do not work alone. Every project calls for collaboration with other professionals, including architects, construction workers, factory foremen, business representatives, government officials, and other kinds of engineers. Being a good team player is vital to the success of every project. If there is discord along the way, mistakes are more likely and projects may not be completed on time or remain within the budget.

Structural engineers come across all kinds of conditions that can affect the functionality of a structure. There may be unfavorable site conditions, constraints from existing structures, or unusual demands from clients. Then there are conditions that were not taken into consideration when an old structure was built that are causing potential problems in the present. In all these cases, structural engineers are responsible for solving the problem. The most successful engineers relish a challenge – it is what makes their job rewarding. The best among them are able to come up with quick and efficient solutions.

There is no room for mistakes in this work. Being meticulous is essential. The slightest error in calculations or design could lead to catastrophic results, such as the collapse of an overpass or failure of an airplane fuselage. Structural engineers pay attention to every detail, especially while reviewing calculations, running computer simulations, and testing models.

Communications skills are particularly important in this field because structural engineers deal with a wide range of individuals on a daily basis. They need to keep open communications going within their team, and mediate conflicts as they come up. They also have to explain complex concepts to a wide range of clients who may be unfamiliar with structural engineering principles. The best structural engineers are able to speak high-level engineering lingo with other engineers and quickly switch to more common language when talking to non-engineers.

Successful structural engineers have a love of learning. Structural engineering projects cross over into a variety of fields, which offers an interesting opportunity to broaden one's knowledge in other areas. Plus, there are new sources of information, methods, materials, and technologies continually adding to the engineering knowledge base. Keeping up with the latest advances can be exciting or it can be a chore. If it feels more like a chore, this probably is not the right career for you.

ATTRACTIVE FEATURES

STRUCTURAL ENGINEERS ENJOY A GREAT DEAL of professional satisfaction. They make an immeasurable contribution to society. Their talents and creativity allow us to travel by air, land, and water through and across any terrain. They help us make good use of crowded urban land while enjoying beautiful views from tall buildings. And perhaps most important of all, they keep us safe from natural forces like earthquakes and hurricanes.

One of the best things about structural engineering work is watching ideas become reality. When a project is finished, there is something there that you can see or touch. It may be an award-winning skyscraper towering in the city skyline, a beautiful new bridge spanning a broad river, or a futuristic aircraft soaring into the sky. There are plenty of small jobs, and they can be just as rewarding when you see them finished.

The work of structural engineers lasts an incredibly long time. Think of the pyramids in Egypt, still standing after thousands of years. The aqueducts of Ancient Rome can be found today throughout Europe. Both are wondrous feats of structural engineering. Today's structural engineers design buildings and bridges to last for more than 100 years. During that time, the structures they design will be used and enjoyed by thousands or even millions of people. It is amazing to think of the long-lasting impact your work will have.

Structural engineers are respected professionals. Their work is not easy and those who work alongside them know it. Construction professionals, including other engineers, recognize what a vital part of every project the structural engineering role is.

The future looks bright for those looking to enter this field. The demand is high and job security is assured for skilled structural engineers. The field also covers a broad spectrum so there are many opportunities to work on different kinds of projects. There is so much demand that you can easily move from job to job and collect signing bonuses along the way. No one thinks twice about seeing multiple job changes on your résumé. In fact, it is expected of anyone who is good at their job. The typical tenure for a structural engineer with an employer is usually not more than five years.

Do you dream of seeing the world? Many structural engineering firms operate on the global level. Within a company, there may be offices operating in a dozen different countries on several different continents. You can move between those offices as new projects come in, or you can act as a consultant yourself and take on clients with projects in the countries you would like to visit. Very often, travel expenses and additional bonuses are paid to those willing to venture far from home.

Structural engineers are well compensated. Getting the necessary education to enter the profession is not easy, but newly graduated structural engineers can expect their first jobs to pay in the neighborhood of $65,000 a year. After a few years of experience (and getting a master's degree), they can step up to a six figure income with the right employer in the right location.

UNATTRACTIVE ASPECTS

THERE IS GREAT RESPONSIBILITY THAT GOES with this work. You cannot afford to make any mistakes because the lives of many people may depend on your calculations. It can be stressful worrying about whether your design will become an engineering failure. The stakes are high and even a tiny miscalculation could lead to a disaster. Confidence in your abilities and maintaining focus will keep you from succumbing to serious anxiety.

The path to becoming qualified to start practicing and getting paid is not an easy one. Engineering degree programs are intense and difficult even for the brightest students. Advanced math and sciences are tough to master, and some students need more than the usual four years to get through it all. It will be even more of a struggle if advanced placement math and physics classes were not successfully completed in high school.

You may think that once you have graduated, the studying is over – it is not. Structural engineers need to be lifelong learners to remain relevant. Advances in technology are particularly vexing. Computers can now solve complex calculations in seconds, relieving you of many hours of tedious work. Computers have changed the entire landscape of structural engineering roles. Those who have kept up with the changes are in demand.

Structural engineers are notorious for moving around

among different companies and different projects. That keeps the work interesting, especially when it involves travel to new places. The downside is licensing issues. There are different licensing requirements in different states and countries that can create an obstacle to starting immediate work in a new location. It can take time, paperwork, and more testing to earn licensure to work in a new region.

EDUCATION AND TRAINING

NEW STRUCTURAL ENGINEERS CAN START OUT with a bachelor's degree. Most undergraduate students major in civil, mechanical, or aerospace engineering with an emphasis on structural engineering. There are also architectural engineering programs that offer structural engineering coursework or are combined with civil engineering programs.

A master's degree is usually required for senior level and research positions, or to join a college faculty. Some situations call for an advanced degree in business, such as an MBA (Master of Business Administration). Many employers, such as engineering consulting services, also prefer candidates with professional certification.

Structural engineers are trained to strict standards due to the seriousness of the safety issues involved in their work. Core subjects include statics and dynamics, materials science, basic conceptual structural design, engineering graphics, and computer aided design. There are classes that cover specific materials, including reinforced concrete, timber, masonry, composites, and structural steel as they relate to general structural design.

Structural analysis is a very important subject and is

therefore often broken into several classes addressing specific aspects, such as structural failure analysis.

Many schools require structural engineering students to take classes in English or communications because so much of the work structural engineers do consists of verbal and written reports.

Starting at the senior level of undergraduate work or in graduate programs, structural engineering coursework becomes more focused on the actual work. Coursework includes subjects like prestressed concrete design, bridge engineering, space frame design for aircraft, and civil and aerospace structure rehabilitation. In graduate school, students can also choose to focus on advanced structural engineering specializations.

A master's degree is not needed for professional licensing as a structural engineer. That is expected to change in the future. Until that happens, new structural engineers should understand that the learning process will continue regardless of any legal requirements. New technology, materials, methods, and government regulations are introduced every day. It is part of the job to participate in ongoing education and training to keep up with advances.

Licensing

New college graduates need to become licensed before working as structural engineers. Every state and the District of Columbia require structural engineers to be licensed. In most states, that means having the same license as a civil engineer. Getting the license involves being tested by the National Council of Examiners for Engineering and Surveying (NCEES). The exam, known as the Engineer in Training (EIT), covers the fundamentals of engineering and is often taken before graduation. A second exam, The Principles and Practice of Engineering

(PE) is required by some states to obtain the Professional Engineer (PE) license. This test is taken after working under the supervision of a licensed engineer (PE). The amount of required experience varies by state, but is typically four years.

Most states do not have a separate structural engineering license. There are, however, several exceptions. Alaska, California, Hawaii, Illinois, Nevada, Oregon, Utah and Washington require more than a civil engineering license to practice structural engineering. The additional license can only be obtained after the engineer has worked a certain amount of time with a civil engineering license. In some states, such as California, that experience earthquakes, there is additional testing related to seismic principles.

Although it is not required, many structural engineers obtain professional qualifications by joining the Structural Engineering Institute, a special branch of the American Society of Civil Engineers.

EARNINGS

THE AVERAGE SALARY FOR ALL STRUCTURAL ENGINEERS in the US is about $85,000 per year. There is a big range though, depending on a variety of factors including work experience, education level, type of employer, and geographic location.

The median annual salary for newly graduated structural engineers with a bachelor's degree is $65,000. It usually ranges between $55,000 and $70,000, but can be as low as $45,000 with some employers in certain locations. Mid-level structural engineers with more than five years of experience and a master's degree can expect somewhere between $80,000 and $110,000. The range is due to a variety of factors. Senior structural engineers with a master's degree and more than 10 years of experience are paid between $115,000 and $150,000.

Most senior-level structural engineers with advanced degrees earn six figure incomes because they are promoted to management positions. Some become independent consultants and/or start their own businesses where they can earn even more.

The most common employers of structural engineers are structural engineering service firms. They offer good pay and exceptional benefits. Depending on the size of the company, pay usually ranges between $80,000 and $100,000. The large corporate firms, which usually do business in multiple countries, do not necessarily pay higher salaries. They are able to attract the best and brightest with an array of perks like signing bonuses, travel allowances, and 401k contributions. Although a smaller number of structural engineers work in aeronautics, this industry pays the highest. Most of its structural engineers earn well into the six figure range. In the public sector, the best pay is at the federal level. Federal employees earn about $95,000 a year on average. State and local government agencies pay about 10 percent less.

The highest-paying jobs for structural engineers are located in major cities, including Washington DC, Los Angeles, and New York. Houston and Seattle are particularly good places to look for work if salary is a motivating factor. That is mainly due to two major industries being located in these cities. Seattle is home to the largest aeronautics companies, and Houston is where a long list of Fortune 500 oil and gas companies are located. Structural engineers are in high demand in both of these industries.

Most structural engineers work full time and therefore, receive full benefits packages including health insurance, PTO (paid time off), and retirement plans. Those who do contract work and move from business to business may need to arrange and pay for their own benefits.

OPPORTUNITIES

THE JOB OUTLOOK FOR STRUCTURAL ENGINEERS continues to be good. Employment of these professionals is projected to grow over 10 percent during the coming decade. That is faster than the average for all occupations. Structural engineers are responsible for creating vital and safe structures like roadways, bridges, buildings, dams, waterways, aircraft, and various industrial and commercial facilities. They stay close to the projects they design until the projects are completed and in use in order to address any problems that may come up. Their critical expertise makes them some of the most in-demand professionals in the design, engineering, and construction management fields.

The current structural engineering industry is under pressure. Simply put, the structural engineering field needs more recruits. The demand is currently outstripping the supply. The field is slowly becoming more attractive to future engineers, but in the meantime, those in the profession are having to cover the gap. Some structural engineers are finding ways to work more efficiently, usually through the use of more advanced technology. Others are putting in longer hours to meet the demands created by a lack of qualified personnel. Structural engineering services are having to negotiate longer schedules and higher fees to compensate for insufficient staff.

The structural engineering industry is swamped and that has employers scrambling to keep their current staff and attract new workers. When demand exceeds supply, prices naturally rise. That is basic economics. Structural engineering salaries are quickly rising as a result. Many engineering firms are so backlogged they are forced to turn away business and accept only the most profitable projects. This is good news for recently graduated structural engineers. Everyone who wants a job should be able to get one. It does not mean, however, that there is no competition for good jobs. As employers pay higher salaries and offer generous signing bonuses to keep fully staffed, they expect their dollars to buy the best and brightest. Applicants will need to demonstrate their value through experience. New structural engineering candidates who gain experience by participating in a co-op program or internship while in college will have the best opportunities.

It is a well-established fact that the current US infrastructure is aging and either falling into disrepair or becoming obsolete. Many state and local governments have deferred funding all projects while the problems continue to grow. The view on the federal level is not much better. Crumbling schools, roads, transit systems, bridges, airports, dams, levees, and other structures desperately need to be repaired or replaced. The country can only postpone the inevitable so long. The delayed projects will have to be completed in order to protect the public and the environment. It is the business of structural engineers to design, upgrade, and monitor the viability and safety of critical infrastructure.

The structural engineering industry, like many other fields, has gone through a high-tech revolution. Advances in software have led to the design of more complex structures and some structural engineering tasks have become automated. Calculation that used to require days of meticulous work are now completed within seconds, and the end results are even better and more accurate. That might sound like structural engineers are becoming obsolete like so many other kinds of workers, but the opposite has turned out to be the case. Now that technology has removed some routine chores, structural engineers can unleash their creativity and focus more on innovation – something that is more interesting than crunching numbers.

GETTING STARTED

STRUCTURAL ENGINEERS ARE IN DEMAND. You can rest assured there is a job out there for you when you graduate, but do be prepared for some competition. The best way to gain an edge when applying for jobs is to obtain some experience in the field. Apply to as many internships as possible. Ideally, you will participate in one that may lead to a job, but the experience itself is valuable. To make sure you receive a good letter of recommendation, be enthusiastic, work hard, and always treat your supervisors as though you are in a professional setting. You should try contacting a local company to ask for work experience. Do not be afraid to ask – you might be surprised by what you can get.

Start looking for your first job at your school's career center. You will find job listings posted in their office, often before they go online. Be sure to sign up for email updates, which most schools offer. Take advantage of other services the career center has to offer, such as mock interviews and résumé writing workshops. It is important that you know how to polish your résumé to highlight your strongest attributes and develop interview skills.

Watch for upcoming career fairs where you can talk with potential employers. The schedules for these events are also posted in the career center. Ask your professors to let you know if they hear of any job openings. Faculty members are typically closely tied to the industry and often know of job openings before they are publicly announced. Search the online job boards on a regular basis. Pay particular attention to job boards and employment agencies that specialize in engineering jobs. There are also recruiters that can help you land that first job.

Become an active member of professional associations like The Institution of Structural Engineers. Participate in local meetings, seminars, and other events that will put you in contact with potential employers.

ASSOCIATIONS

- **The Structural Engineering Institute**
www.asce.org/sei

- **The National Council of Structural Engineers Associations**
www.ncsea.com

- **American Society for Engineering Education**
https://www.asee.org

- National Council of Examiners for Engineering and Surveying (NCEES)
https://ncees.org

- National Society of Professional Engineers
https://www.nspe.org

PERIODICALS

- Structure
http://www.structuremag.org

- Civil + Structural Engineer
https://csengineermag.com

WEBSITES

- Technology Student Association
www.tsaweb.org

- Engineering Education Service Center
www.engineeringedu.com/store

Copyright 2019
Institute For Career Research   CHICAGO
CAREERS INTERNET DATABASE
www.careers-internet.org

www.ingramcontent.com/pod-product-compliance
Lightning Source LLC
Chambersburg PA
CBHW071200220526
45468CB00003B/1100